THE SUN

will belong to every one of us

Manifesto

NEW WORLD ORDER

Mehmet Kılıç

Translated from the German by Marie Weber.

Mehmet Kılıç
The sun will belong to every one of us
Manifesto "New World Order".
1. Auflage

Mehmet Kılıç

E-Mail: lwn.mtp@gmx.de
Internet: www.mehmetkilic.com
Youtube: "The sun will belong to every one of us"
Herstellung und Verlag:
BoD – Books on Demand, Norderstedt

ISBN 978-3-7534-3103-1

MIX
Papier aus verantwortungsvollen Quellen
Paper from responsible sources
FSC® C105338

PREFACE

The prevailing social system, with its competitive spirit and philosophy, compels people and society to possess more and more, to constantly need to win, to become ever more powerful in order to have dominance.

This system exerts such great pressure on the individual that people will go to any lengths to fight each other endlessly in their endeavour to have the upper hand.

Throughout this brutal struggle, however, we people are destroying our basic living conditions, the only and unique common habitat of all living things and are leading the way towards the end of life on our planet.

With his "New World Order" under the motto "The sun will belong to every one of us" Mehmet Kilic advocates the elimination of all thoughts and actions whose causes and effects pose a threat to life on Earth.

The author presents concrete proposals to secure life on Earth and create a humane model of society.

His proposals contain fundamental changes and differences in content and method to the ideas and thoughts of many philosophers and scientists such as John Locke, Montesquieu, Immanuel Kant, Karl Marx, Albert Einstein, Hans Küng, Otfried Höffe, Zhao Tinyang et al, on the subject of "World Peace" and "New World Order".

The author interprets life on Earth according to his philosophy of the one unifying whole, whose core contains such universal principals as solidarity and equality, on which his manifesto New World Order is based.

With his manifesto Mehmet Kilic appeals to mankind to stop all actions which endanger life on Earth and to build a humane model of society for us all.

THE SUN

will belong to every one of us

Manifesto

NEW WORLD ORDER

Mehmet Kılıç

Translated from the German by Marie Weber.

Heartfelt thanks goes to my dear friends Marie Weber and Joanne Lowe for their linguistic support.

INTRODUCTION

Dear readers,

The sun will belong to every one of us.
I am pleased to present to you my manifesto "New World Order".
Life on our planet is unique and unparalleled. All of the plants, all of the animals and all men are worthy of our love, appreciation and protection. It is Mother Nature – air, water and land – which is especially worthy of protection.

Yet how we treat Mother Nature is actually very disquieting!

We have already sawed through a good part of the green branch on which we are all sitting. When we fall, not only shall we seven thousand million people fall, but together with us, our cats and roses.

If on the one hand we continue to fight one another and on the other hand saw off the branch on which we are sitting, we shall experience our very last breath as a hitherto unheard of scream – the last sign of this senselessness and madness, all of which will not be recorded in history books.

As Founding Father and Chairman of the registered peace initiative Hand in Hand Bad Kreuznach, I have been working for a more dignified life in this world for years. I would like to offer a resumé of my philosophical-political thoughts of many years on the theme of world peace and present the "New World Order" to a world-wide audience.

In the event that you, too, are worried about life on this earth of ours and are searching for a way out, my manifesto could be of interest to you.

PART I

MY PROBLEM

I am very worried about the future of life on our planet.

Reasons

- Peace and security are disappearing all over the world, both for individuals and for societies.
- The basic requirements for the existence of all living things are being destroyed.

Causes for this development

- The present world order forces man and society to compete: to win, to become richer and richer, stronger and stronger and to conquer more and more in order to have control.
- This craving is encouraged to the extent that worse consequences such as endless wars, exploitation, poverty and other painful repercussions are the result. This is "MADNESS"!
- At the same time the system leads to the destruction of the shared habitat of all living things and, with it, the conditions necessary for existence. Thus the system is leading to the "BITTER END" of life on earth!

Inddition it is clear to me that:

1. The prevailing social system stands in contradiction to common sense.

2. One cannot expect the mind-set which turns work and effort into money, profit and power, to secure the happiness of mankind.

3. One can also not expect the same mentality to prevent the abuse of mankind's common habitat, to put an end to the destruction of basic living conditions and to bring about a durable protection of life on our planet.

4. Mankind is unhappy! The threat is terrifying! The MADNESS becomes more insane every day! The BITTER END is approaching incredibly fast. There is no more time to lose!

5. Neither one person, nor one nation, nor one state alone is responsible for this situation. The responsibility lies with the prevailing world order.

6. No one should close his eyes in the face of this MADNESS or in the face of the rapidly approaching BITTER END!

7. The only power capable of stopping the MADNESS and averting the BITTER END is mankind itself.

CONCLUSION

The existing world order can no longer be upheld. It has become a problem for mankind – for which a solution must definitely and urgently be found.

SUGGESTED SOLUTION

In order to solve this problem I propose establishing a new world order.

THE NEW WORLD ORDER

The objective of the new world order will be to permanently secure the happiness of mankind on earth.

PART II

Question 1:
WHAT WILL THE NEW WORLD ORDER LOOK LIKE?

Three main pillars will serve as a basis for the new world order; together they will support, strengthen and protect each other:

Pillar I: The Philosophy of the One Unifying Whole
Pillar II: Child-rearing Practices and the Educational System
Pillar III: The Universal Basic Rules of Life

Now to the individual pillars:

Pillar I
THE PHILOSOPHY OF ONE UNIFYING WHOLE

In order to illustrate the Philosophy of the One Unifying Whole, I would like to take you along on a short journey! Please close your eyes and imagine the following:

THE UNIVERSE AND ONE UNIFYING WHOLE

- The unending universe forms one unifying whole together with its countless stars, constellations and galaxies, which themselves form their own unities and entireties.
- The Milky Way, as an indispensable part of the universe, forms one unifying whole.

- The solar system, as an inseparable and indispensable part of the Milky Way, forms one unifying whole.
- The earth, as an inseparable part of the solar system, is one unifying whole.
- The world of living things, as an inseparable part of the earth, is one unifying whole.
- Mankind, as an inseparable part of the world of living things, is one unifying whole.

THE INDIVIDUAL AND ONE UNIFYING WHOLE

The individual, which we call "man", embodies one unifying whole with his entire being and is an indispensable part of mankind.

- The body of an individual consists of organs, which consist of parts and countless cells.
- Each organ, each part and each cell forms its own unifying whole.
- Each organ is arranged differently and has a different shape and a different job and function from other organs; it can also work independently of other organs.
- If one of the organs is missing or fails to function, the physical unity is incomplete and the wholeness does not exist.
- If one of the organs is separated from the physical whole, that is the end of its life, just as a cell also dies when it is removed from an organ.

- A man can only function as an individual and lead his life when all of his organs do their job together in harmony.

Picture this for a better understanding:

Peter is sitting in the kitchen. His nose has picked up a smell. He realizes that it is the smell of something burning. He turns his head toward the cooker. What does he see? A piece of newspaper is burning on the cooker. As a result Peter jumps up, runs to the cooker and extinguishes the flames.

- The organ that detects the smell is Peter's nose.
- The organ that sees the flames is not Peter's stomach, it's his eyes.
- The organs that enable him to get to the cooker are not his hands, but his feet.

What would have happened,

- if Peter's nose had not been able to smell anything?
- if Peter's eyes had not been able to see the fire?
- if Peter's hands had not been capable of putting out the fire?

Conclusion:

With all of its organs a body forms one unifying whole and fulfills its functions, *when* all of the organs work together in harmony.

What do you think: Can we apply this principle to life in society?

SOCIETY AND ONE UNIFYING WHOLE

Life in society begins when at least two people come together due to something which unites them at that moment.

Let us say, for example, that a young woman and a young man find each other attractive. They make the decision to live their lives together.

- At the outset of the decision it is clear, who will play what role in a particular area of life.
- In other areas it becomes clear only after a time, who assumes the responsibility for what.
- In those areas in which the woman has more competence, the man will play a supportive role.
- In those areas in which the man is stronger, the woman will play a secondary role.

Conclusion:

This young couple could form the smallest social unifying whole. By acting in harmony, the two could establish and lead a life together.

Another example:

A family would like to build its own house. Can this family build a house from the foundation to the roof without outside help?

When building a house, a considerable number of people from various occupational groups would have to be involved, including architects, brick-layers, roofers etc. People from the aforementioned groups come together and form one unifying whole. The members of this unifying whole act in combination and build the house.

We can apply the manner in which the builders are active to all areas and all sectors of life in society.

MANKIND AND ONE UNIFYING WHOLE

- The individual is made up of organs and forms a unifying whole.
- Social organs are made up of single individuals and they in turn form a unifying whole.
- Societies are made up of social organs and each society forms a unifying whole.
- Mankind consists of societies and forms the complete unifying whole.

Thus mankind forms one unifying whole, just as the universe does with its countless stars, constellations and galaxies.

Mankind represents a unifying whole to the same extent that the individual does, for example, Peter, who is made up of organs, which in turn are made up of parts and countless cells.

Mankind is just as much a unifying whole as a social organ, for example, the building and construction industry, which is made up of individuals with different skills and abilities, such as architects, bricklayers, roofers etc.

Now I shall summarize everything:

THE CONSEQUENCE FOR THE INDIVIDUAL

Every individual, every person who has completely understood the Philosophy of the One Unifying Whole will discover a path that leads him to his self.

On this inward journey he will:

1. discover and get to know the indescribable beauty and riches of his inner world and
2. develop a sense of how important and how precious he is.

He will begin

- to accept himself as he is and to no longer compare himself with other people,
- to respect himself and to love himself more and more,
- to radiate outward the beauty of his inner world and his increasing zest for life,
- to see and to approach and to deal with all people as if every one of them were "he himself",
- to place his respect for Mother Nature on a true, healthy and stable foundation.

Thus every individual who has internalized the Philosophy of the One Unifying Whole, will find the true SELF, therefore HIMSELF and will comprehend the meaning of universal life.

THE CONSEQUENCE FOR SOCIETY

Once the Philosophy of the One Unifying Whole has been internalized, a process of change will begin in the personality of the individual! A process of change, which will soon be noticeable in the social life of all mankind.

As a result of that –

- People will no longer measure each other's worth on the basis of their different skills and capabilities or on the basis of their standing in society.
- They will regard no one more or less worthy of respect than themselves.

- They will consider their work to be no more or less useful than the work of others.
- People will realize that all groups, communities, societies and mankind form one unifying whole.

THE CONSEQUENCE FOR MANKIND

- We can apply the behavior of an individual, for example of Peter, to the behaviour of a social organ, for example the building and construction industry.
- We can transfer the behavior of a social organ, for example the building and construction industry, to the whole society, for example that of Germany.
- We can transfer the behavior of a society, for example that of Germany, to human society as a whole, that is to say, to the "body of mankind".

The most important principles and values of the One Unifying Whole:

1. A common bond
   We belong together; like the organs of a human body, the staff of a hospital, the crew of an airplane, the workers in a factory ...
2. Equality
   We are all equal;
   it is without question that all men are equal, and

it is without question that positions and jobs are equal.

I now create a synthesis of "solidarity" and "equality" and state:

We are all for us all!

Since all the individuals, who make up social organs, societies and all of mankind will thoroughly internalize the Philosophy of the One Unifying Whole, they will organize and lead their life in the spirit of this philosophy. We are all for us all!

For this reason they will devote what they all have in common

- their intellect,
- their conscience
- and their consciousness

  to all aspects of their personal and social life, in order to protect and secure life on our planet, individual happiness and the happiness of mankind.

PART III

Pillar II

CHILD-REARING PRACTICES AND THE EDUCATIONAL SYSTEM

The objective is for individuals to be trained and educated as new generations that will organize and lead their personal and social lives based on the Philosophy of the One Unifying Whole.

In order to attain this objective, the establishment of the New World Order will be launched with a sustainable program called “Development from the cradle to the grave”.

Now I shall attempt to present the “Development from the cradle to the grave”.

DEVELOPMENT OF THE EXISTING FRAMEWORK

The entire earth is provided with basic conditions which take the changes into account that arise from the new system of child-rearing and education and the new world order.

In addition “Centers for Child-rearing, Education and Lifestyle” are set up around the world with the aim of

- demonstrating the natural style and conduct of life,

- utilizing the earth's natural resources in a meaningful way and
- eliminating the causes of mass exodus and mass transports.

TEACHER TRAINING

Teacher candidates must have adequate capabilities for this occupation and really want the job.

During training the particular capabilities of all the candidates are developed in such a way as to equip them with a varied and sound set of skills.

The candidates receive the appropriate pedagogic training for specific subject areas and age groups.

CONTENT

1. CONVEYING THE PHILOSOPHY OF THE ONE UNIFYING WHOLE

In the cradle each baby is first given the finest basic values of the new philosophy of life, much like it is given its mother's milk.

In school the children learn to internalize and live the philosophy of the One Unifying Whole.

2. FOSTERING PERSONAL TALENTS

All the natural abilities of each and every child are explored and identified as early as possible and developed to a high level of competency.
The child has the ability to use all of his or her skills in his or her private and social life.

3. PERSONALITY DEVELOPMENT

- The child is accompanied and supported with attentiveness, care and concern,
- He will get to know himself and his inner world and learn to love and have regard for himself,
- The child will respect other people and learn to have regard for and protect them as he does himself,
- He will learn how to apply his personal "values" in the social context.

4. PREPARATION FOR PRIVATE AND SOCIAL LIFE

- The young people are explicitly prepared both for their private as well as their social life in accordance with the Philosophy of the One Unifying Whole.
- They are raised and educated to become healthy individuals with highly developed skills.

- They are not trained as specialists for existing occupations, but rather as stable individuals with a well-developed sense of responsibility.
- They are put in a position to not only shape their own life successfully and with pleasure, but also play a constructive, creative and active role in the life of society.

Pillar III
UNIVERSAL BASIC RULES OF LIFE

The Universal Basic Rules of Life are to be understood as the Constitution of the World State. Since they must first be drawn up by the Founding Parliament and adopted by the World Parliament of Elected Representatives, I do not wish to go into further detail at this moment.

PART IV

Question 2:
HOW WILL THE NEW WORLD ORDER BE ESTABLISHED?

WORLD STATE

In order to be able to establish the new world order – which will certainly not happen so easily and so quickly – mankind requires a strong, stable and dependable organization. This will be the World State.

Question 3:
WHAT WILL THE WORLD STATE LOOK LIKE?

A) THE ORGANIZATIONAL PHILOSOPHY OF THE WORLD STATE

The organization of the World State demonstrates by example that mankind, just as a healthy human being, doesn't live contrary to its own organs and cells, but lives for the whole; thus the synthesis of the values and principles of the Philosophy of One Unifying Whole is made evident: Everyone is for everyone else!

DEMOCRACY AND ITS FUNCTIONING

Democracy is reorganized or optimized:

Ideas, thoughts and actions which

- lead to the destruction of Mother Nature,

- endanger and make human life more difficult and
- could jeopardize human dignity

will have no place in the new understanding of democracy.

All of the ideas, thoughts, forms of behavior and activities

- which are in accord with Mother Nature and with all living creatures, especially the health and dignity of man, and
- which are accepted by the common human intellect and conscience and confirmed by a common sense of responsibility,

will be counted among the unlimited and fundamental democratic freedoms.

DIVISION OF RESPONSIBILITY

The principle of the separation of powers gives way to the principle of the "division of responsibility".

Reasons:

1. Owing to the fact that each individual has a strong sense of responsibility, there will be no need for criticism, use of influence or outside control mechanisms.
2. Everyone will fulfill his task in the service of the community with the same sense of responsibility which he would apply in his personal and social life.

ELECTIONS

In the election districts all areas of life are newly defined and determined.

"Areas of life" refer to those sectors in which people generally work for their living, for example in agriculture, a factory, the field of education etc.

ELIGIBILITY AND TRANSPARENCY

Everyone is accepted as a candidate for an office if he can prove that he is qualified to represent an area of life in a parliament.

The candidates are elected by the voters specifically for a committee, in order that they represent a particular area of life in the parliament concerned.

Therefore it becomes more and more clear, exactly who voted for whom and for which purpose, and who was elected by whom for which task.

FLEXIBLE RENEWABILITY

The principle of "flexible renewability" allows for a change to be made at any time by those elected. This guarantees that the functionality and efficiency of the World State is maintained at the optimal level.

POLITICAL PARTIES

The political parties lose their basis for existence for two important reasons:

1. It goes without saying that people will devote their common intellect, conscience and consciousness for common goals to organizing life in society.

2. Candidates will be elected not because of their membership in a party or organization, but because of their qualifications.

B) ORGANIZATIONAL STRUCTURE OF THE WORLD STATE

THE ORGANIZATION OF CIVILIAN LIFE

The People for the People

The organization of the service provider "The People for the People" will emerge from the people, with the people and for the people and serve all people free of charge.

In the People's Information Centers citizens will get quick, dependable and understandable answers to their simple questions regarding "What" and "Where".

In the People's Advisory Centers citizens will receive detailed and authoritative information from experts for projects such as 'How do I build a house?'

In People's Assistance Centers trained members of staff will help people solve their problems. If, for example, someone wants to build a house, he will be assisted beginning with the planning and calculating of costs all the way up to the layout of the garden.

In addition to staff members, philosophers, guides and mediators will also offer their services in these centers.

Philosophers will work with the personnel in all of the service centers and administrative units, so that services will be carried out in accordance with the Philosophy of the One Unifying Whole.

Guides will make certain that services in the People's Advisory Centers and People's Assistance Centers concur with the "Rules" of the parliaments.

The mediators will limit their work to the People's Assistance Centers and bring about an agreement between parties of differing opinions when it comes to a particular subject.

THE ORGANIZATION OF THE STATE

The organizational structure of the World State will consist of:

1. Local administrative units
   (like today's local governments and city administrations)

2. Regional administrative units
   (like the states in Germany or countries like
   France with a centralist government)
3. Continental administrative systems
4. The World Parliament of Elected Representatives

The administrative systems will deal with those themes which belong to their areas of responsibility and jurisdiction.

PART V

Question 4:
WHAT WILL THE WORLD STATE DO?

A) IMMEDIATE MEASURES

In a first step the World State will implement a series of measures.

Immediate measure 1:

Stop the MADNESS ! Avert the BITTER END !

That includes

- stopping the destruction of nature,
- disbanding all military facilities, destroying all weapons and ending all wars.
- feeding all of the hungry,
- offering shelter to all of the homeless,
- putting an end to human trafficking and sex slavery
- .....

Immediate measure 2:

EXPANSION OF INFRASTRUCTURES

The infrastructures in all areas of life around the world are reorganized in accordance with the values and principles of the Philosophy of the One Unifying Whole.

Immediate measure 3:

PROTECT LIFE ON EARTH

In order to safeguard life on this earth, the World State, along with its citizens, will
1. put an end to the destruction of Mother Nature, and
2. eliminate the causes of poverty and its consequences by introducing a "minimum standard of living".

B) ESTABLISHING THE NEW WORLD ORDER

In a second step the World State, together with its citizens, will begin to establish the new world order.

All areas of life around the world will be redefined and reorganized in the light of:

- the Philosophy of the One Unifying Whole,
- the new System of Child-rearing and Education and
- the Universal Basic Rules of Life

The period of transition between the old and the new system will be especially influenced and shaped by:

1. consideration, empathy and understanding for others and for existing values on the one hand, and
2. resolve and determination when establishing the new world order on the other hand.

REORGANIZATION OF ADDITIONAL AREAS OF LIFE

Some examples:

REORGANIZATION OF THE JUDICIAL SYSTEM

The judicial system will be reorganized for the following reasons:

1. It has been proven that national laws are incompatible with universal reason; the infeasibility of these laws has proven to be real.
2. No one, neither an individual, nor society, nor the state will be in a privileged position.
3. The new system will produce neither wrong-doing nor culprits.
4. The causes for wrong-doing will disappear by themselves.
5. The probability of being unjustly treated and of having to defend oneself will simply disappear.
6. The new world view will permanently include universal justice as an achievement of mankind.

RULES INSTEAD OF LAWS

Living together in the new world will be regulated by means of RULES, and not by means of laws!

Parliaments will not pass any laws, which, if violated, would then require the application of criminal law.

They will develop “basic rules”, which arise from man’s natural needs and which individuals, society and the state abide by.

The “basic rules” with key functions will

- be held at a minimum,
- be formulated in simple sentences always in the first person and
- be available for learning in the world language as well as in the native languages of all citizens.

If it is a question of abiding by the rules for a certain activity, for example building a house, one can consult the rules at “The People for the People”.

UNDECIDED CASES

Guides and mediators from the People’s Assistance Centers will intervene in undecided cases.

DISSOLUTION OF THE MILITARY

The military's reasons for existing will disappear by themselves, since the previous demand for its activity will no longer exist; for example:

- the conquest of territories, resources and riches that belong to others,
- the defense of a country or a people against the attacks of others etc.

DISSOLUTION OF THE POLICE FORCE

There will no longer be a need for the existence of the police, since the previous duties involving support, protection and surveillance will be unnecessary after the dissolution of the judicial system and the founding of "The People for the People".

DISSOLUTION OF SECRET SERVICE ORGANIZATIONS

Because the conditions that have driven individuals and societies to "Madness" no longer exist, it will prove senseless to collect secret information about others.

As a result the secret service organizations will be dissolved.

DISSOLUTION OF NATIONAL BORDERS

Based on the realization that

- the artificial division of mother nature is meaningless and
- all living creatures belong to the earth and not the other way around,

the borders between countries will be done away with.

PART VI

Question 5:
HOW WILL THE WORLD STATE BE FOUNDED?

initiative groups

- As soon as people realize that the prevailing social system is at an end and it is urgently necessary to take action, these people will form initiative groups everywhere.
  In these initiative groups people thoroughly internalize the Philosophy of the One Unifying Whole.
- They adopt my manifesto "New World Order", exchange information and discuss everything.
- The initiative groups organize themselves globally and form networks among themselves.
- They begin a worldwide information campaign by informing people
  - about the intolerability of this madness and the dimensions of the danger and
  - about the manifesto "New World Order".

- As soon as the information campaign has reached its goal, the initiative groups will call on the international community to form a Parliament for the World State.

FOUNDING PARLIAMENT

- In response to the appeal of the initiative parliaments national states will carry out elections for representatives to the Founding Parliament.

- The Founding Parliament
  - begins with its work in a suitable location,
  - formulates the Universal Basic Rules of Life,
  - conducts a worldwide information campaign with the focus on the Universal Basic Rules of Life and
  - following that, conducts a survey of citizens on this theme.

- As soon as it is evident that both the information campaign and the citizens' survey have been successful, the Founding Parliament invites the states of the world to participate in the foundation of the World Parliament of Elected Representatives.

WORLD PARLIAMENT
OF ELECTED REPRESENTATIVES

- The Founding Parliament invites the people to elect representatives from their countries to the World Parliament of Elected Representatives.
- The World Parliament of Elected Representatives
  - takes up its work,
  - adopts the Universal Basic Rules of Life and
  - elects the President of the World State.

- Thereupon the world population celebrates the founding of the World State.

***

Dear readers,

I have briefly presented to you my concerns regarding the very unique life on our planet and my suggestion for a solution to this problem of mankind.

I hope that the urgency of taking action and the practicality of my manifesto have become perfectly clear and that my dearest wish may be fulfilled.

The situation is serious!

We have no time to lose!
We are standing at a fork in the road.
It is time to make a historical decision!

It is time to begin!
It is time for us to unite!

Let us put an end to the Madness, eliminate the danger and prevent the
bitter end!

First and foremost this is a call to all those who love themselves, who love roses, carnations and tulips, who love cats, dogs and pigeons!

First and foremost this is a call to all those who respect and appreciate Mother Nature, people and mankind!
And I myself see it as my duty to devote all my energies and ability to this pressing undertaking.

I wish us all much success!
I give you my sincere thanks!
The sun will belong to every one of us.
Good-bye and take care.